What Plants Produce

BY MARTIN J. GUTNIK

ILLUSTRATED BY SAM SHIROMANI

CHILDRENS PRESS, CHICAGO

Library of Congress Cataloging in Publication Data

Gutnik, Martin J
  What plants produce.

  SUMMARY: Presents experiments which demonstrate that plants produce oxygen and carbohydrates.
1. Photosynthesis—Experiments—Juvenile literature.
[1. Photosynthesis—Experiments] I. Shiromani, Sam.
II. Title.
QK882.G92   581.1'3342   75-37619
ISBN 0-516-00527-8

Copyright © 1976 by Regensteiner Publishing Enterprises, Inc.
All rights reserved. Published simultaneously in Canada.
Printed in the United States of America.

1 2 3 4 5 6 7 8 9 10 11 12 R 78 77 76

To Aunt Sunny, Uncle Bill, and Cousin Sheldon,
whose faith and support helped me to write this book.

CONTENTS

FOREWORD 7
DO PLANTS REALLY MAKE OXYGEN? 11
DO PLANTS MAKE SUGAR AND STARCH? 27
WORDS YOU SHOULD KNOW 39
ACKNOWLEDGMENTS 43
ABOUT THE AUTHOR AND ARTIST 45

FOREWORD

Photosynthesis is the process by which green plants make food. It comes from two words: *photo,* which means light, and *synthesis,* which means putting together.

Green plants put things together to make food, They use light to help them. Green plants are the only living things on earth that can make their own food.

All animals and human beings need green plants for food. Without green plants there could be no food.

Green plants also make *oxygen.* Oxygen is the gas we breathe. The plants make oxygen at the same time they make food. If green plants did not make oxygen, we could not live. We would not have enough oxygen to breathe.

This book will help you to learn about what a green plant makes. Green plants make two things: food and oxygen.

The food green plants make is *sugar.* It is called *glucose.* The plants change this sugar to other things. It changes the sugar to starch. We use starch for food and energy. The plant uses starch for food and energy.

Green plants also make oxygen. They make oxygen while they are making food.

By doing the experiments in this book you will learn how plants make oxygen. You will learn about the food plants make.

DO PLANTS REALLY MAKE OXYGEN?

Bill's teacher told him a lot about plants. He told him that plants make sugar. This sugar is the food that all living things needs. The word for the process by which plants make food is *photosynthesis.*

Bill's teacher also told Bill that while a plant is making food it also makes oxygen. Oxygen is the gas we breathe. We need it to live.

If plants did not make oxygen there would not be enough for us to breathe. Bill wondered if plants really do make oxygen. He decided to do an experiment to see if this was really true.

Things Bill needed:
1. An elodea plant. This is a water plant. It can be bought at any fish store or pet store.
2. A funnel.
3. A fish bowl or any deep bowl.
4. A test tube.
5. Baking soda.
6. Water.
7. One match (to be used only by an adult).
8. A toothpick.
9. A lamp.
10. Small pitcher or sink.

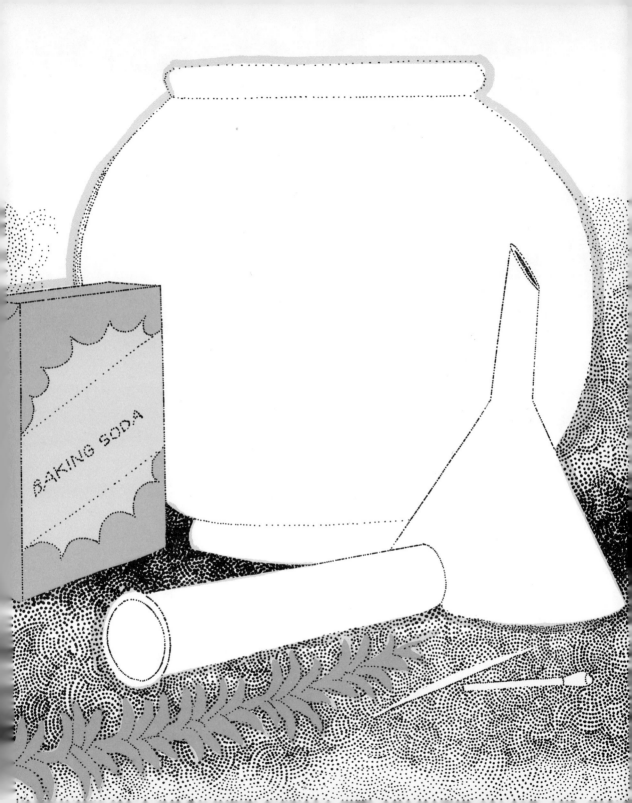

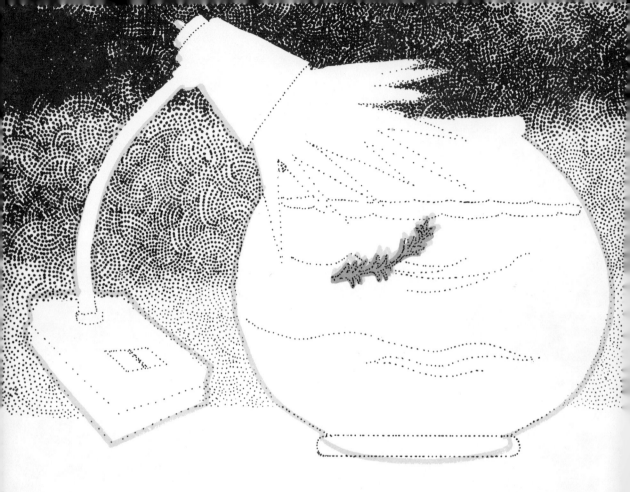

What Bill Did

Bill filled a fish bowl with water. He put his elodea plant in the fish bowl. He let the plant float on the water. He put a lighted lamp over the fish bowl. Plants need light to make food.

Then Bill held a funnel upside down. He put the upside-down funnel into the fish bowl. He covered the elodea plant with the wide end of the funnel. He made sure that the whole funnel was covered with water.

Bill had a good reason for putting the plant under the funnel. If the elodea made oxygen, he wanted the oxygen to go out the small end of the funnel.

Bill's teacher told him that oxygen is a gas. A gas will bubble in water. It will go to the top of the water.

Bill had to catch the oxygen that his plant would make. To do this he used a test tube. He filled the test tube to the top with water.

Then Bill put
his thumb over the
mouth of the test tube.

He turned the
test tube upside down.

Bill put the upside-down test tube under the water in the fish bowl. He held it above the small end of the funnel. Then he removed his thumb under the water. He did this so that all the water would stay in the test tube.

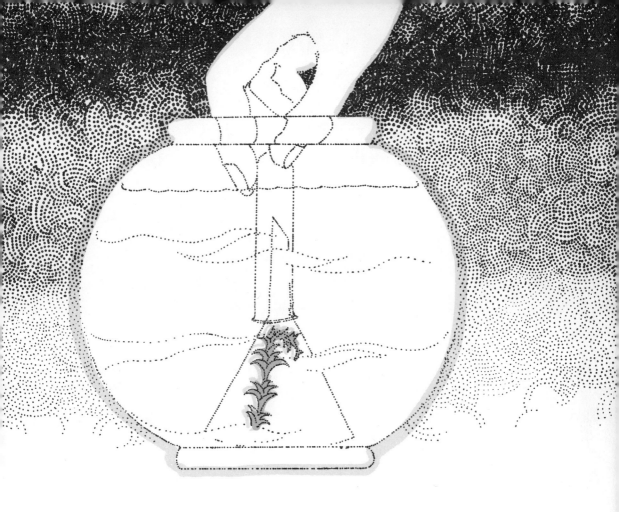

Then Bill placed the test tube of water over the small end of the funnel.

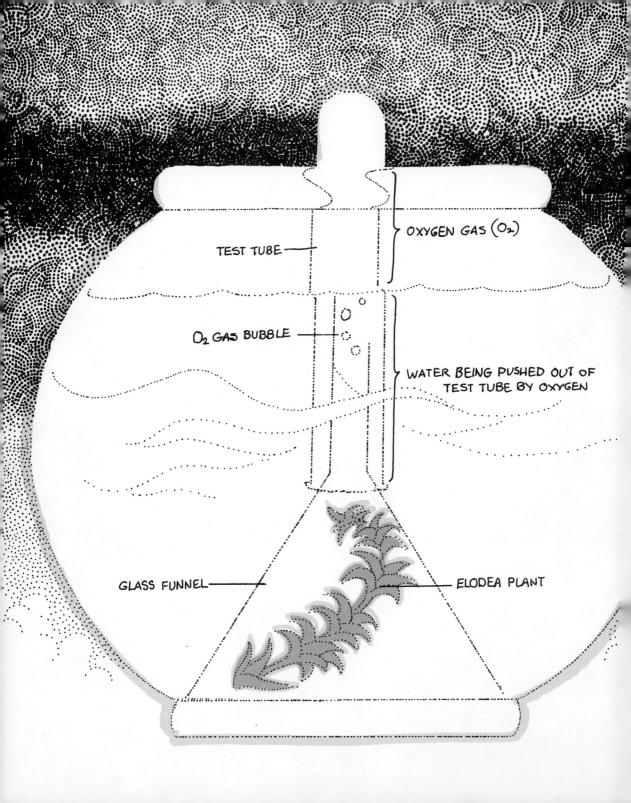

Bill's teacher had told him that the oxygen gas would bubble in the water. The bubbles would float to the top. If Bill's elodea plant were making oxygen, then the oxygen should bubble in the water.

Bill had put the funnel over the plant so the oxygen bubbles would have to go through the funnel. They would then go into the test tube. The bubbles would push the water out of the test tube. When all the water was out of the test tube it would be full of oxygen.

Bill waited until the test tube was full of oxygen. This took about two days. For some plants it might take longer. For other plants it might not take that long.

When the test tube was empty of water and full of oxygen, **Bill** had to take it out of the fish bowl. He had to test it to see if it really did have oxygen in it.

First Bill put his hand in the fish bowl. He slowly lifted the test tube off the funnel. **He was very careful not to lift the test tube out of the water.**

Bill put his thumb over the mouth of the test tube. He did this so that the oxygen could not get out of the tube. He lifted the test tube out of the water. He still kept his thumb over the mouth of the test tube.

Now Bill was ready to test for oxygen. **He needed to have an adult help him. He could not do the test by himself.** Bill asked his teacher to light a match. He asked him to use the match to light a toothpick on fire.

Bill's teacher let the toothpick burn for about five seconds. After five seconds he blew it out. The teacher quickly put the glowing end of the toothpick into the test tube.

Bill had to move his thumb at just the time his teacher blew out the toothpick.

What Did Bill See?

When Bill's teacher put the glowing toothpick into the test tube, the toothpick caught fire again.

Bill's teacher told him that oxygen helps fire burn. When the toothpick caught fire, Bill knew that there was oxygen in the test tube. The oxygen in the test tube had come from the plant.

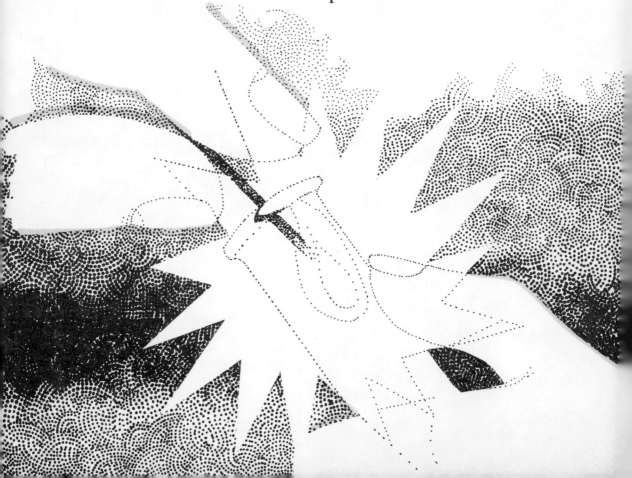

DO PLANTS MAKE SUGAR AND STARCH?

Bill's teacher told him that plants make sugar. He told him that plants use this sugar for food.

Green plants change sugar to starch. Sugar and starch are carbohydrates. This is a food used for energy. Plants use this energy to grow and make more food.

Bill wanted to see if plants really do make starch. He decided to do an experiment to find out.

Things Bill needed:
1. A potato.
2. A butter knife or table knife.
3. A medicine dropper.
4. Iodine.

What Bill Did

Bill's teacher told him that iodine could be used to test for starch. When iodine is put on starch it turns a very dark blue. This tells you that starch is there.

Bill cut a potato in half. He used a potato because his teacher told him that potatoes were starchy.

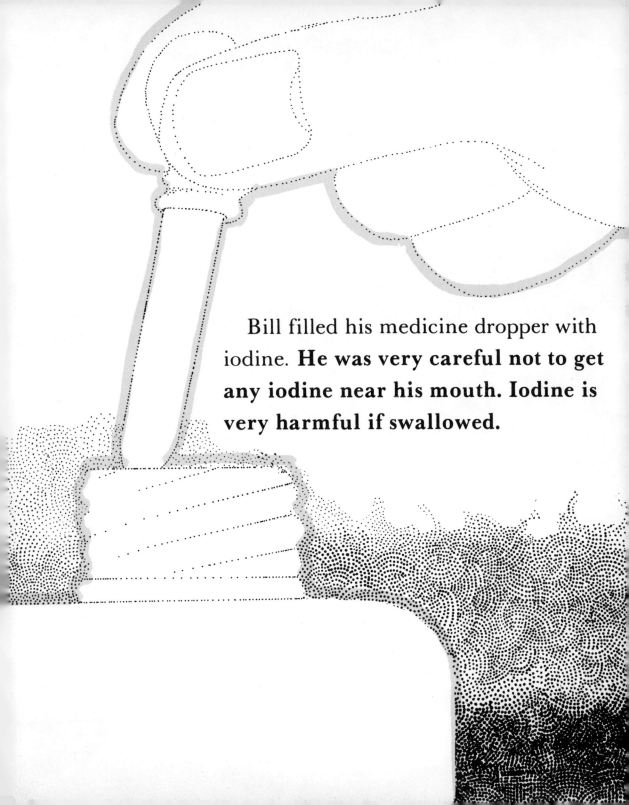

Bill filled his medicine dropper with iodine. **He was very careful not to get any iodine near his mouth. Iodine is very harmful if swallowed.**

Then Bill put the iodine from his medicine dropper onto the potato. He waited to see what would happen.

What Did Bill See?

Bill saw that the potato turned a dark blue. This meant that the potato was starch. Iodine turns dark blue when it touches starch.

Bill now knew that a potato plant makes starch. Bill thought that if a potato plant makes starch then other plants must make starch, too.

To be sure that a potato plant was not the only plant that makes starch, Bill decided to do another experiment.

Things Bill needed:
1. A kernel of corn.
2. Cornstarch.
3. Iodine.
4. A medicine dropper.
5. Two small dishes.
6. A teaspoon.
7. A small cup.
8. A butter knife or a table knife.

What Bill Did

Bill put a kernel of corn into a small cup of water. He let it soak in the water overnight. He did this to soften it. He was going to cut it open and test for starch.

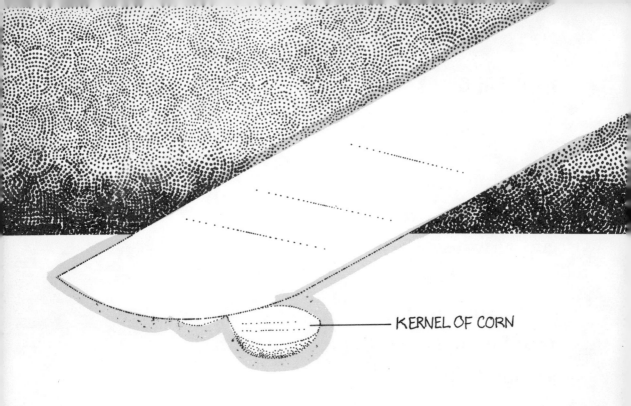

KERNEL OF CORN

The next day Bill took the kernel of corn out of the water. He cut it in half with his butter knife. He put the halves into a small dish.

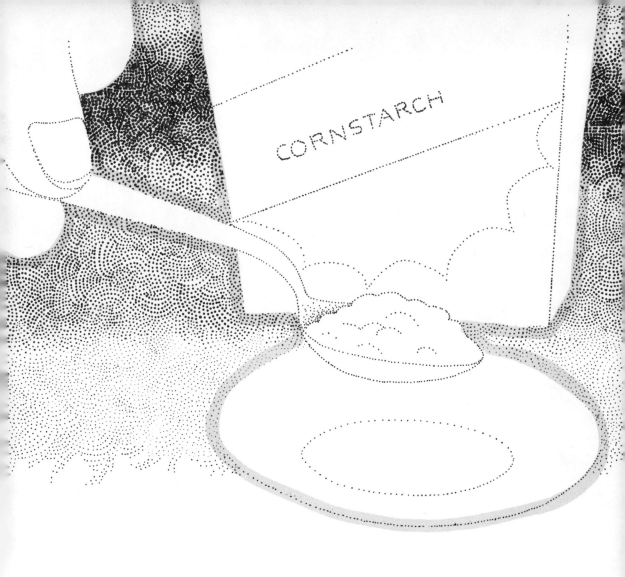

Bill then put a teaspoon of cornstarch into another small dish. He had learned that cornstarch comes from corn. He wanted to test it for starch.

Bill filled his medicine dropper with iodine. He put the iodine onto the two halves of the corn kernel.

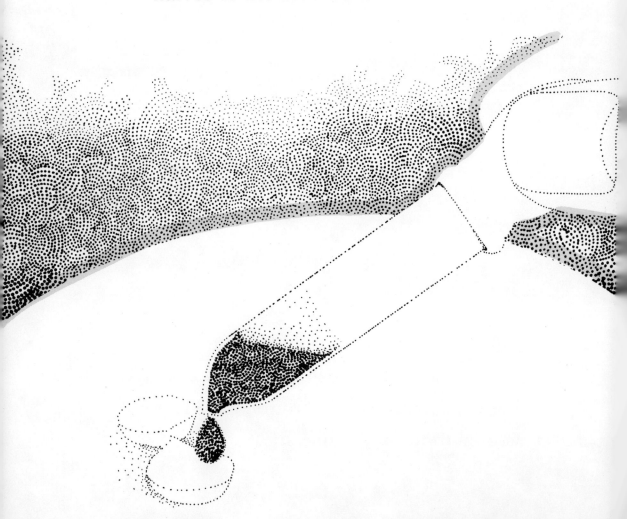

Bill filled his medicine dropper again. This time he put the iodine on the corn starch. He watched what happened to the kernel and to the cornstarch.

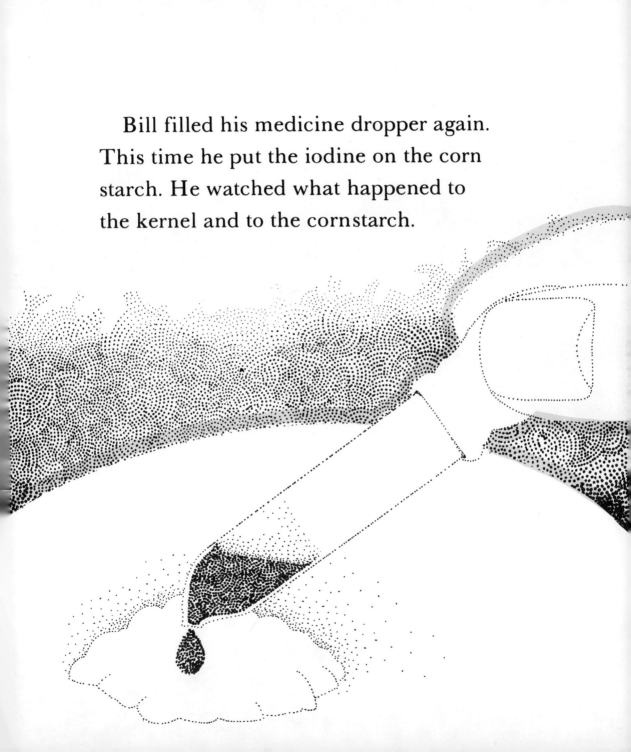

What Did Bill See?

Bill saw that both the kernel of corn and the cornstarch turned a dark blue. This showed him that they both had starch.

Bill had now learned that other plants besides potatoes make starch. His teacher told him that all plants make starch. They need starch for energy. Energy helps them make food. Energy helps them make oxygen.

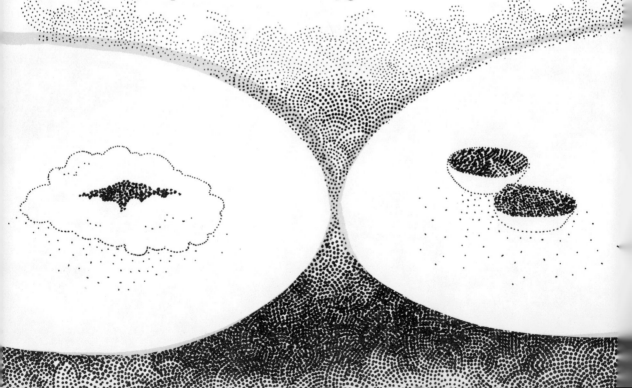

WORDS YOU SHOULD KNOW

Carbohydrate (KAHR boh HI drate) A form of food. Usually sugar or starch.

Carbon dioxide (CO_2) (kahr bohn dye AHK side) A gas that green plants use to help them make food. Plants get this gas from the air.

Chlorophyll (KLOHR oh fill) A green coloring in plants. It absorbs (sucks in) the sun's energy and uses it to help the plant make food. Chlorophyll is the only thing on earth that can absorb the sun's energy to make food.

Chloroplast (KLOHR oh plast) A plant tissue that holds chlorophyll.

Iodine solution (EYE oh dine suh LOO shun) Iodine mixed in water: one teaspoonful of iodine per quart of water. It can be used as a test for starch. When iodine solution is mixed with starch it turns a dark blue.

Light energy (lite ehn uhr gee) The energy plants use to make food.

Oxygen (OHKS ih jen) A gas we breathe. Green plants make this gas when they make food.

Petiole (PET ee ohl) The leaf stem. The part of the leaf that is attached to the stem of the plant.

Phloem (FLOW ehm) Tubes (veins) in a plant that carry the sugar that the plant makes.

Photosynthesis (foh toh SIN thuh sihs) Made up of two words: *photo,* which means light, and *synthesis,* which means putting together. A plant puts water and carbon dioxide together. It uses light as its helper. It also uses chlorophyll as a helper. When these things are put together they make sugar and oxygen.

Raw material (rahw mah TEER ee ahl) A substance in its natural state.

Solution (suh LOO shun) A substance that is mixed with water.

Starch (stahrch) A food made by plants. Plants make sugar and change the sugar to starch.

Stomata (STOH mah tah) Holes in leaves of plants; like the pores in skin.

Sugar (glucose) (SHUHG ahr) (GLOO kohs) A simple sugar produced by plants in photosynthesis.

Tissue (TISH yew) A group of cells that look the same and do a certain job.

Transpiration (TRAN spih ray shun) Occurs when plants lose water from their leaves.

Water (H_2O) (WAH ter) A liquid that plants use as a raw material to make food.

Xylem (ZYE lehm) Tubes (veins) that carry water and food to the plant.

ACKNOWLEDGMENTS

Peter Berg—Sixth grade student, Atwater School, Shorewood, Wisconsin

Margaret (Mrs. Bruce) Berg—Illustrator, science aide, Atwater School, Shorewood, Wisconsin

John Tranetski—Graduate student, University of Wisconsin, Milwaukee

Forest Stearns—Professor of Ecology, University of Wisconsin, Milwaukee

Timothy Stearns—Sixth grade student, Atwater School, Shorewood, Wisconsin

About the Author:

Martin Gutnik, an innovative elementary school science teacher, lives with his wife and two young children in Milwaukee, Wisconsin. His three latest books for Childrens Press, *How Plants Are Made, How Plants Make Food,* and *What Plants Produce,* are based on a series of experiments on photosynthesis that he created for his students to perform in the classroom. His first three *First Experiments in Science & Nature* books were based on the same type of experiments in the field of ecology and pollution. Through Mr. Gutnik spends much of his time helping his students learn to build ecosystems, develop film, and dissect frogs, his life is not totally surrounded by science. He also collects records from the fifties and enjoys singing folk songs as he accompanies himself on the guitar, which he taught himself to play.

About the Artist:

Born in India, Sam Shiromani has made Chicago his home. A dropout from the world of advertising, he devotes most of his time to free-lancing art and photography.